Stamford foundry Co.

Cooking and Heating Stoves Ranges Hot-Air Furnaces

Laundry Stoves Confectioners' Stoves Caboose Ranges, etc.

Stamford foundry Co.

Cooking and Heating Stoves Ranges Hot-Air Furnaces
Laundry Stoves Confectioners' Stoves Caboose Ranges, etc.

ISBN/EAN: 9783744799232

Printed in Europe, USA, Canada, Australia, Japan

Cover: Foto ©berggeist007 / pixelio.de

More available books at **www.hansebooks.com**

THE STAMFORD

FOUNDRY COMPANY

ESTABLISHED 1830   INCORPORATED 1869

COOKING AND HEATING STOVES RANGES

HOT-AIR FURNACES

LAUNDRY STOVES CONFECTIONERS' STOVES

CABOOSE RANGES

ETC ETC

SALESROOMS AND WORKS

STAMFORD CONN

1892-3

THE ALLEY-ALLEN PRESS NEW YORK

A S a rule, cuts give a very good idea of a range or stove or furnace. If the cut is accompanied with a clear description of the article shown, a dealer can easily form an opinion of it. He will certainly derive no help from extravagant praise bestowed on the goods by the manufacturer.

This book is published for the use of dealers, and useless matter is left out.

THE STAMFORD

FOUNDRY COMPANY.

NORMANDIE RANGE.

In all that constitutes a first-class leading range, the Normandie is complete.

Its ovens are larger than the standard size; its flues are large and so arranged that it works quickly and thoroughly.

It has large top, extension shelf, reversible pipe collar, Read's improved check-damper, wide broiling door, illuminated fire door, spring knobs on draft-slides, tin-lined oven door, polished flanges, nickel mountings.

Duplex grate, deep end hearth with large square ash pan.

The flues are large; castings smooth, heavy, and well fitted. Oven door has patent attachment for opening with the foot.

Each size is made in all the forms represented in the following cuts.

Read's patent check-damper is used with all Normandie Ranges, unless plain collar is specially ordered.

Towel-rail is always furnished.

Instead of the usual wood-fixtures, a cast-iron pan can, if desired, be furnished, which fits the bottom of the fire-box, and can be put in or removed in a moment, without disturbing the brick or coal-grate. This very convenient arrangement, it is believed, is found in no other line of ranges.

NORMANDIE RANGE.

	Size of Oven	Number of Cover Holes	Size of Cover Holes
No. 77	18½ x 18½ x 11	6	7-inch
No. 78	18½ x 18½ x 11	6	8-inch
No. 88	21 x 21 x 12	6	8-inch
No. 89	21 x 21 x 12	6	9-inch

The same range body can be used on closet, base, or plain feet without any alteration — a great convenience to dealers.

Plain range, as shown above, will be shipped, unless some other form is specially ordered.

Water-fronts furnished for each size, if ordered.

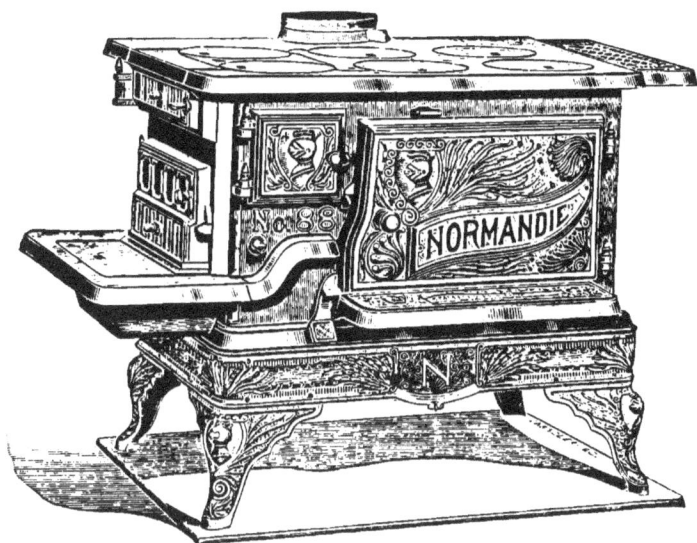

NORMANDIE RANGE

With Ornamental Base.

	Size of Oven	Number of Cover Holes	Size of Cover Holes
No. 77	18½ x 18½ x 11	6	7-inch
No. 78	18¼ x 18¼ x 11	6	8-inch
No. 88	. . 21 x 21 x 12	6	8-inch
No. 89	. 21 x 21 x 12	6	9-inch

The same range body can be used on closet, base, or plain feet, without any alteration — a great convenience to dealers.

Water-fronts furnished for each size, if ordered.

NORMANDIE RANGE

With Reservoir.

	Size of Oven	Number of Cover Holes	Size of Cover Holes
No. 77	18½ x 18½ x 11	6	7-inch
No. 78	18½ x 18½ x 11	6	8-inch
No. 88	21 x 21 x 12	6	8-inch
No. 89	21 x 21 x 12	6	9-inch

The same range body can be used on closet, base, or plain feet, without any alteration — a great convenience to dealers.

NORMANDIE RANGE

With Base and Reservoir.

	Size of Oven	Number of Cover Holes	Size of Cover Holes
No. 77	18¼ x 18¼ x 11	6	7-inch
No. 78	18½ x 18½ x 11	6	8-inch
No. 88	21 x 21 x 12	6	8-inch
No. 89	21 x 21 x 12	6	9-inch

The same range body can be used on closet, base, or plain feet, without any alteration — a great convenience to dealers.

NORMANDIE RANGE

With Low Closet.

	Size of Oven	Number of Cover Holes	Size of Cover Holes
No. 77	18¼ x 18¼ x 11	6	7-inch
No. 78	18¼ x 18¼ x 11	6	8-inch
No. 88	21 x 21 x 12	6	8-inch
No. 89	21 x 21 x 12	6	9-inch

The same range body can be used on closet, base, or plain feet, without any alteration — a great convenience to dealers.

Water-fronts furnished for each size, if ordered.

NORMANDIE RANGE

With Reservoir, Low Closet, and High Shelf.

			Size of Oven	Number of Cover Holes	Size of Cover Holes
No. 77	.	. . .	18½ x 18½ x 11	6	7-inch
No. 78	.	. . .	18½ x 18½ x 11	6	8-inch
No. 88	.	. . .	21 x 21 x 12	6	8-inch
No. 89		. . .	21 x 21 x 12	6	9-inch

The above shelf is removable, and can be used with any form of the range.

The same range body can be used on closet, base, or plain feet, without any alteration—a great convenience to dealers.

DUCHESS RANGE.

A modern and well appointed range of the highest grade.

Some of its most important features are : the large top with extension shelf ; reversible pipe-collar ; wide broiling door ; deep end hearth with large, square-framed ash pan ; pedal oven-door opener ; wide polished flanges ; handsome nickel trimmings ; smooth castings, and careful mounting ; and the construction of fire-box so that either duplex grate or flat grate can be used.

Either half or both halves of the duplex grate can be removed and replaced without disturbing the bricks or water-front, as can also the flat grate.

Read's patent check-damper is furnished in place of plain collar, if ordered, without extra charge.

Towel-rail is always furnished.

Wood-fixtures can be furnished for all sizes.

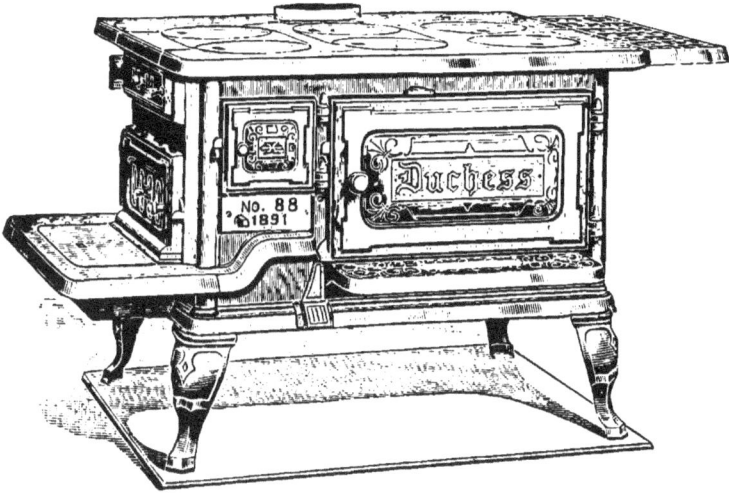

DUCHESS RANGE.

		Size of Oven	Number of Cover Holes	Size of Cover Holes
No. 77	Duplex Grate	18 x 18 x 10	6	7-inch
No. 78	" "	18 x 18 x 10	6	8-inch
No. 88	" "	. 20 x 20 x 11	6	8-inch
No. 99	" "	20 x 20 x 11	6	9-inch
No. 2	Flat Grate	18 x 18 x 10	6	7-inch
No. 2½	" "	18 x 18 x 10	6	8-inch
No. 3	" "	20 x 20 x 11	6	8-inch
No. 4	" "	20 x 20 x 11	6	9-inch

The same range body can be used on closet, base, or plain feet, without any alteration — a great convenience to dealers.

Plain range, as shown above, will be shipped, unless some other form is specially ordered.

Water-fronts furnished for each size, if ordered.

DUCHESS RANGE

With Ornamental Base.

		Size of Oven	Number of Cover Holes	Size of Cover Holes
No. 77	Duplex Grate . .	18 x 18 x 10	6	7-inch
No. 78	" " .	18 x 18 x 10	6	8-inch
No. 88	" "	20 x 20 x 11	6	8-inch
No. 99	" "	20 x 20 x 11	6	9-inch
No. 2	Flat Grate . .	18 x 18 x 10	6	7-inch
No. 2½	" " . .	18 x 18 x 10	6	8-inch
No. 3	" " .	20 x 20 x 11	6	8-inch
No. 4	" "	20 x 20 x 11	6	9-inch

The same range body can be used on closet, base, or plain feet, without any alteration—a great convenience to dealers.

Water-fronts furnished for each size, if ordered.

DUCHESS RANGE

With Reservoir.

Sizes made with reservoir, No. 88, No. 99, No. 3, No. 4.

		Size of Oven	Number of Cover Holes	Size of Cover Holes
No. 88	Duplex Grate	20 x 20 x 11	6	8-inch
No. 99	" "	20 x 20 x 11	6	9-inch
No. 3	Flat Grate	20 x 20 x 11	6	8-inch
No. 4	" "	20 x 20 x 11	6	9-inch

The same range body can be used on closet, base, or plain feet, without any alteration — a great convenience to dealers.

DUCHESS RANGE

With Base and Reservoir.

Sizes made in this combination, No. 88, No. 99, No. 3, No. 4.

			Size of Oven	Number of Cover Holes	Size of Cover Holes
No. 88	Duplex Grate	.	20 X 20 X 11	6	8-inch
No. 99	"	"	20 X 20 X 11	6	9-inch
No. 3	Flat Grate		20 X 20 X 11	6	8-inch
No. 4	"	"	.	20 X 20 X 11	6	9-inch

The same range body can be used on closet. base, or plain feet. without any alteration — a great convenience to dealers.

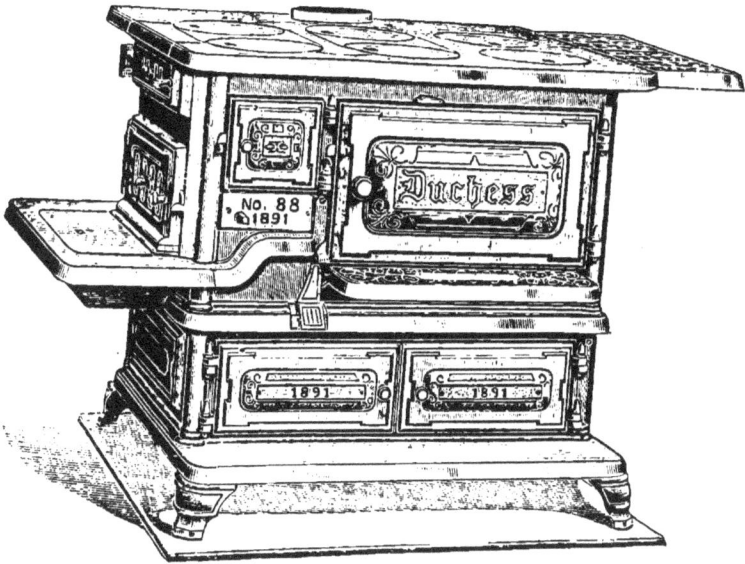

DUCHESS RANGE

With Low Closet.

		Size of Oven	Number of Cover Holes	Size of Cover Holes
No. 77	Duplex Grate	18 x 18 x 10	6	7-inch
No. 78	" "	18 x 18 x 10	6	8-inch
No. 88	" "	20 x 20 x 11	6	8-inch
No. 99	" "	20 x 20 x 11	6	9-inch
No. 2	Flat Grate	18 x 18 x 10	6	7-inch
No. 2½	" "	18 x 18 x 10	6	8-inch
No. 3	" "	20 x 20 x 11	6	8-inch
No. 4	" "	20 x 20 x 11	6	9-inch

The same range body can be used on closet, base, or plain feet, without any alteration — a great convenience to dealers.

Water-fronts furnished for each size, if ordered.

DUCHESS RANGE

With Reservoir, Low Closet, and High Shelf.

Sizes made in this combination, No. 88, No. 3, No. 99, No. 4.

The above shelf is removable, and can be used on any form of the range.

The same range body can be used on closet, base, or plain feet, without any alteration — a great convenience to dealers.

DUCHESS RANGE

With Base, Reservoir, and Elevated Closet.

DUCHESS RANGE

With Removable End Hearth.

For situations where it is desired, in summer, to put range into a fire-place which is too small to admit the form of range shown in preceding cuts.

The internal construction of the range is the same, the difference being in the arrangement of the ash pit.

This form of the Duchess is also made with low closet, or reservoir, or both, as shown in preceding cuts.

For numbers and dimensions, see previous pages (15 to 21).

Wood-fixtures can be furnished for all sizes.

Water-fronts furnished for each size, if ordered.

Memoranda.

OUR OWN RANGE.

An attractive and popular range, equipped with the best of modern appliances for convenience and perfect working.

Has large top with extension shelf; polished flanges; nickel trimmings; tin-lined oven door; pedal oven-door opener; square, framed ash pan.

Fire-box is so constructed that either duplex grate or flat grate can be used.

No. 88 and No. 33 fitted with wood-fixtures, if desired.

OUR OWN RANGE.

		Size of Oven	Number of Cover Holes	Size of Cover Holes
No. 77	Duplex Grate . .	17½ x 18 x 9	6	7-inch
No. 88	" "	. 19½ x 19¾ x 10	6	8-inch
No. 22	Flat Grate	. 17½ x 18 x 9	6	7-inch
No. 33	" " . .	. 19½ x 19¾ x 10	6	8-inch

Plain range, as shown above, will be shipped, unless some other form is specially ordered.

The same range body can be used on closet, base, or plain feet, without any alteration — a great convenience to dealers.

Water-fronts furnished for No. 88 and No. 33, if ordered.

OUR OWN RANGE

With Ornamental Base.

		Size of Oven	Number of Cover Holes	Size of Cover Holes
No. 77	Duplex Grate	. 17½ x 18 x 9	6	7-inch
No. 88	" "	. 19½ x 19¾ x 10	6	8-inch
No. 22	Flat Grate	. 17½ x 18 x 9	6	7-inch
No. 33	" "	. 19½ x 19¾ x 10	6	8-inch

The same range body can be used on closet, base, or plain feet, without any alteration — a great convenience to dealers.

Water-fronts furnished for No. 88 and No. 33, if ordered.

OUR OWN RANGE

With Reservoir.

Sizes made with reservoir, No. 88 and No. 33.

		Size of Oven	Number of Cover Holes	Size of Cover Holes
No. 88	Duplex Grate	19½ x 19¾ x 10	6	8-inch
No. 33	Flat Grate	19½ x 19¾ x 10	6	8-inch

The same range body can be used on closet, base, or plain feet, without any alteration — a great convenience to dealers.

OUR OWN RANGE

With Base and Reservoir.

Sizes made in this combination, No. 88 and No. 33.

		Size of Oven	Number of Cover Holes	Size of Cover Holes
No. 88	Duplex Grate .	19½ x 19⅞ x 10	6	8-inch
No. 33	Flat Grate .	19½ x 19⅞ x 10	6	8-inch

The same range body can be used on closet, base, or plain feet, without any alteration — a great convenience to dealers.

OUR OWN RANGE

With Low Closet.

		Size of Oven	Number of Cover Holes	Size of Cover Holes
No. 77	Duplex Grate .	. 17½ x 18 x 9	6	7-inch
No. 88	" "	. 19½ x 19¾ x 10	6	8-inch
No. 22	Flat Grate .	17½ x 18 x 9	6	7-inch
No. 33	" "	19½ x 19¾ x 10	6	8-inch

The same range body can be used on closet, base, or plain feet, without any alteration — a great convenience to dealers.

Water-fronts furnished for No. 88 and No. 33, if ordered.

OUR OWN RANGE

With Reservoir, Low Closet, and High Shelf.

Sizes made in this combination, No. 88 and No. 33.

		Size of Oven	Number of Cover Holes	Size of Cover Holes
No. 88	Duplex Grate	19½ x 19⅞ x 10	6	8-inch
No. 33	Flat Grate	19½ x 19⅞ x 10	6	8-inch

The same range body can be used on closet, base, or plain feet, without any alteration — a great convenience to dealers.

The above shelf is removable, and can be used on any form of the range.

Memoranda.

OUR OWN RANGE

With Removable End Hearth.

For situations where it is desired, in summer, to put range into a fire-place which is too small to admit the form of range shown in preceding cuts.

The internal construction of the range is the same, the difference being in the arrangement of the ash pit.

This form of the Our Own is also made with low closet, or reservoir, or both, as shown in preceding cuts.

For numbers and dimensions, see previous pages (25 to 30).

Wood-fixtures can be furnished for No. 33 and No. 88.

Water-fronts furnished for No. 88 and No. 33, if ordered.

Memoranda.

FAIRVIEW RANGE

With Reservoir and Low Closet.

A right-hand fire-box range, with flat grate.

Grate can be removed and replaced without disturbing anything else — a great convenience in repairing.

This range, with the exception of difference in grate, is the same as the Countess, for description and different styles of which, see following pages.

Water-front furnished for No. 8, if ordered.

COUNTESS RANGE.

Duplex grate.   Right-hand fire-box.

	Size of Oven	Number of Cover Holes	Size of Cover Holes
No. 7	16 x 16 x 10	6	7-inch
No. 8	18 x 18 x 11	6	8-inch

Right-hand fire-box renders this handsome range preferable for many housekeepers, and a necessity in cases where the arrangement of kitchen doors, sinks, hot-water boiler, etc., would otherwise interfere.

Water-fronts furnished for No. 8, if ordered.

COUNTESS RANGE

With Reservoir.

Duplex grate.   Right-hand fire-box.

	Size of Oven	Number of Cover Holes	Size of Cover Holes
No. 8	. 18 x 18 x 11	6	8-inch

COUNTESS RANGE

With Low Closet.

Duplex grate.   Right-hand fire-box.

		Size of Oven	Number of Cover Holes	Size of Cover Holes
No. 7	.	16 x 16 x 10	6	7-inch
No. 8	.	18 x 18 x 11	6	8-inch

Water-fronts furnished for No. 8, if ordered.

COUNTESS RANGE

With Reservoir and Low Closet.

Duplex grate.   Right-hand fire-box.

Size of Oven	Number of Cover Holes	Size of Cover Holes
18 x 18 x 11	6	8-inch

No. 8

Memoranda.

UNCLE NICK RANGE.

For those who require a range of medium size, there is nothing to be more highly recommended than the Uncle Nick.

The ornamentation is artistic and pleasing.

The castings are heavy and smooth.

The range is well put together.

It works with remarkable quickness, even with a small or dull fire.

An important feature of this range is that the halves of the duplex grate can be removed and replaced without disturbing bricks or water-front.

		Size of Oven	Number of Cover Holes	Size of Cover Holes
No. 11	Duplex Grate . .	16 x 16 x 9	6	7-inch
No. 12	" " . .	17½ x 18 x 10	6	8-inch
No. 4	Flat Grate . . .	16 x 16 x 9	6	7-inch
No. 5	" " .	17½ x 18 x 10	6	8-inch

Water-fronts furnished for No. 12 and No. 5, if ordered.

UNCLE NICK RANGE

With Ornamental Base.

		Size of Oven	Number of Cover Holes	Size of Cover Holes
No. 11	Duplex Grate	. 16 x 16 x 9	6	7-inch
No. 12	" "	. 17½ x 18 x 10	6	8-inch
No. 4	Flat Grate .	. 16 x 16 x 9	6	7-inch
No. 5	" "	. 17½ x 18 x 10	6	8-inch

Water-fronts furnished for No. 12 and No. 5, if ordered.

ADVOCATE RANGE.

A plain, substantial range, well made, and possessing excellent working qualities.

Both sizes of the Advocate are also made with low closet; and No. 8 is made with reservoir, or with reservoir and low closet.

This range is, in all its forms, fitted with combined check and anti-dust damper, by means of which the fire is easily controlled, and dust is prevented from escaping into the room when grate is disturbed.

Grate can be removed and replaced without disturbing anything else — a great convenience in repairing.

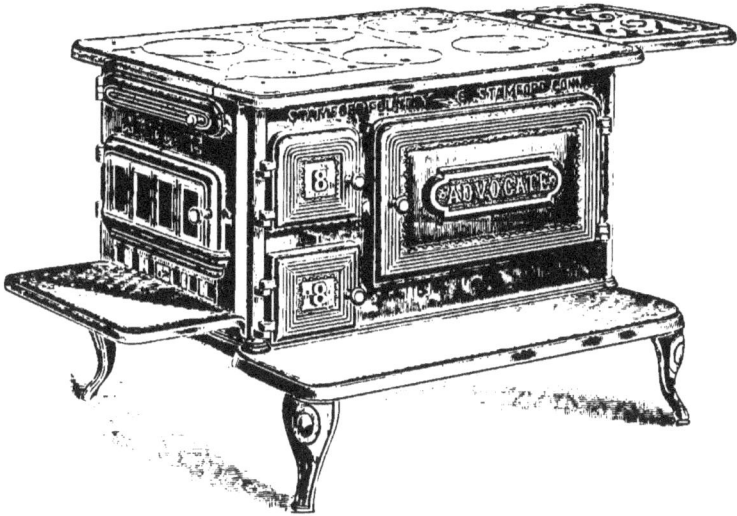

ADVOCATE RANGE.

Anti-Clinker Grate.

	Size of Oven	Number of Cover Holes	Size of Cover Holes
No. 7	. 18 x 18 x 10	6	7-inch
No. 8	. 20 x 20 x 11	6	8-inch

Wood-fixtures can be furnished for both sizes.
Water-fronts furnished for each size, if ordered.

COMET A RANGE.

The history of the Comet line of ranges is one of long and continued success, from the introduction of the original Comet Range up to the present day.

Although the range has never been made in large and highly ornamented forms, it has maintained its popularity, and is likely to for years to come.

The Comet was one of the earliest ranges in this country.

The Comet A is the fourth and latest form, and possesses all of the good points of the original range, with the advantage of added improvements.

It has broiler door with check-draft, extension top shelf, automatic oven shelf, mica door, and nickel trimmings. Pipe can be taken out top or back.

Grate can be removed and replaced without disturbing anything else — a great convenience in repairing.

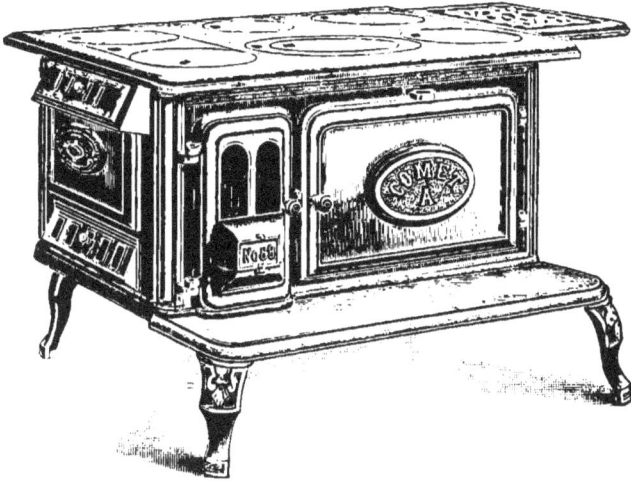

COMET A RANGE.

No. 77	16 x 16 x 9
No. 88	18 x 18 x 10

No. 77 has five 7-inch holes and one 4½-inch hole, also ring to make one 9-inch hole.

No. 88 has five 8-inch holes and one 5½-inch hole, also ring to make one 10-inch hole.

No. 88 can be fitted with a water-front, by being mounted with a special end.

Wood-fixtures can be furnished for both sizes.

COMET A RANGE

With Low Closet.

	Size of Oven
No. 77	16 x 16 x 9
No. 88	18 x 18 x 10

No. 77 has five 7-inch holes and one 4½-inch hole, also ring to make one 9-inch hole.

No. 88 has five 8-inch holes and one 5½-inch hole, also ring to make one 10-inch hole.

No. 88 can be fitted with a water-front, by being mounted with a special end.

Wood-fixtures can be furnished for both sizes.

NEW COMET RANGE.

		Size of Oven
No. 6	. .	12 x 14 x 8
No. 7	.	13½ x 15½ x 8¾
No. 8		18 x 16 x 10

No. 6 has five 6-inch holes, with ring to make one 8-inch hole. No. 7 has five 7-inch holes, with ring to make one 9-inch hole. No. 8 has five 8-inch holes, with ring to make one 10-inch hole.

With its drop oven door and heavy, durable fire plates, covers, etc., the New Comet is well adapted to use on vessels, and has had a large sale for that purpose.

The No. 8 New Comet can be fitted with water-front, by being mounted with special end.

Wood-fixtures can be furnished for No. 7 and No. 8.

HECTOR RANGE.

	Size of Oven	Number of Cover Holes	Size of Cover Holes
No. 57	15 x 16 x 9	5	7-inch
No. 58	18 x 18 x 9¾	5	8-inch

The Hector Ranges, and the Ajax Ranges shown on the opposite page, form a series of fine appearing, good working, low priced ranges. The ovens and flues are large, the top flue, especially, being made deep to accommodate any hollow-ware in general use.

AJAX RANGE.

Two Holes.

Size of Oven	Size of Cover Holes	With Ring to increase one hole to
13 x 12 x 8	7-inch	8-inch
14 x 13 x 8½	8-inch	9-inch
15 x 14 x 9	9-inch	10-inch

Three Holes.

Size of Oven	Size of Cover Holes
13 x 12 x 8	6-inch
14 x 13 x 8½	7-inch
15 x 14 x 9	8-inch

Five Holes.

EXTRA COOK.

Coal or Wood.

	Size of Oven
No. 7 .	17 x 14 x 9
No. 8 .	19 x 16½ x 10

A heavy, well made, quick-working cook stove. Has nickel knobs, brilliant illuminating front, clinker-freeing grate, extra large deep ash-hearth, heavy covers, and strong, well braced cross-pieces.

ORBIT.

Coal or Wood.

	Size of Oven
No. 6	15 x 12½ x 8½
No. 7	17 x 14 x 9
No. 8	19 x 16½ x 10
No. 9	21 x 18 x 11

Perfect working, cheap cook. Has open front grate and dumping bottom grate, extra large deep ash-hearth, heavy covers, and strong, well braced cross-pieces. Quick in operation. Well made and heavy throughout.

No. 128 STAMFORD RANGE.

Made both for brick-setting and in portable form, with either right hand or left hand fire-box.

It is one of the very best of this type of range.

Has triangular revolving grates.

Grates can be taken out by loosening one bolt, without disturbing frame, or brick, or water-front.

Oven is of unusual hight.

Oven damper can be removed by loosening one bolt, without disturbing any other part of the range.

All six cover-holes have direct heat, when damper is open.

No. 128 STAMFORD BRICK SET RANGE
With Low Closet.

Made either right hand or left hand fire-box. Furnished with or without water-front.

No. 128 STAMFORD BRICK SET RANGE
With Low Closet and Elevated Shelf.

Made either right hand or left hand fire-box. Furnished with or without water-front.

No. 128 STAMFORD PORTABLE RANGE
With Low Closet and Elevated Shelf.

Made either right hand or left hand fire-box. Range can be had without Elevated Shelf, if desired. Furnished with or without water-front.

THE WINDSOR.

For Hotels, Restaurants, and Boarding Houses.

Portable or Brick Set.

A most economical, durable, and quick acting range.

It has two ovens, with automatic shelves, and sliding dampers for regulating the heat of either oven; also direct-draft damper and dust damper, both working from front of range.

It has double anti-clinker grate, dumping either to the front or side; a sifting grate which dumps; and a capacious ash pan.

The feed and broiling door is large, and all of the fire doors are easily removed, making an excellent place for roasting.

The water-back is large, and a powerful heater.

The range has two spacious warming closets, besides a plate shelf extending the full length of the range.

A nickel guard rail and polished edges improve its appearance, and it is made extra heavy to stand rough usage.

The range can be sent without the water-back and shelf, if desired.

AN IMPORTANT FEATURE. Oven sides against fire-box, if ever burned out, can be easily removed and replaced by loosening three bolts. Any one can do it, and save the expense, otherwise necessary, of taking the range to pieces.

THE WINDSOR.

Two Sizes.

	Length of Range	Hight of Range	Width of Top	Size of Ovens	Number of Cover Holes	Size of Cover Holes
No. 89	4 ft. 6 in.	2 ft. 7 in.	2 ft. 6 in.	18 x 22 x 13½	8	9-inch
No. 9	5 ft.	2 ft. 7 in.	2 ft. 6 in.	20 x 22 x 13½	8	9-inch

THE CORT.

The Cort is highly praised, both by dealers and by persons using it.

It is suited for hard work, and is sure to do good work.

Its two ovens are fitted with automatic shelves, and with dampers for regulating the heat of either oven independently of the other. Dust damper, operated from the front of the range, removes the annoyance of dust and ashes in the kitchen.

The range has double anti-clinker grate, dumping either to front or side; or, if preferred, triangular revolving grate. It has a dumping ash-sifting grate. Under each oven is a warming closet.

AN IMPORTANT FEATURE. Oven sides against fire-box, if ever burned out, can be easily taken out and replaced, by loosening three bolts. Any one can do it, and save the expense, otherwise necessary, of taking the range to pieces.

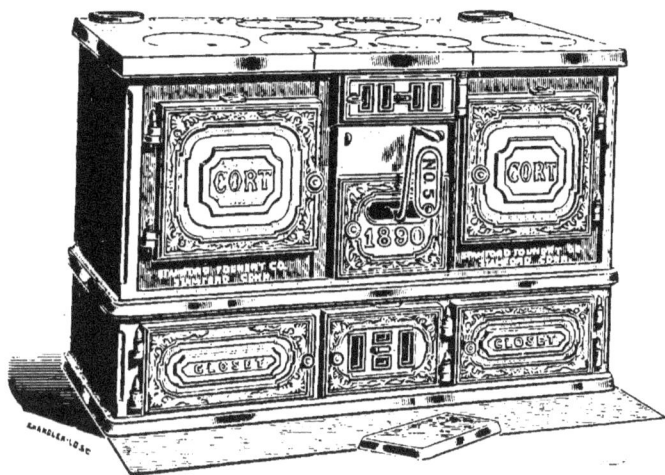

THE CORT.

Three Sizes.

This range is equally efficient as a portable range, or for brick setting.

FLAT GRATE.

	Length of Top	Hight of Range	Width of Top	Size of Ovens	Number of Cover Holes	Size of Cover Holes
No. 2	3 ft. 5 in.	2 ft. 2 in.	22½ in.	11 x 18 x 9½	6	7-inch
No. 2½	3 ft. 7 in.	2 ft. 2 in.	2 ft.	11 x 18 x 9½	6	8-inch
No. 3	3 ft. 11 in.	2 ft. 4 in.	2 ft.	13 x 19 x 11	6	8-inch

TRIANGULAR GRATE.

	Length of Top	Hight of Range	Width of Top	Size of Ovens	Number of Cover Holes	Size of Cover Holes
No. 4	3 ft. 5 in.	2 ft. 2 in.	22½ in.	11 x 18 x 9½	6	7-inch
No. 4½	3 ft. 7 in.	2 ft. 2 in.	2 ft.	11 x 18 x 9½	6	8-inch
No. 5	3 ft. 11 in.	2 ft. 4 in.	2 ft.	13 x 19 x 11	6	8-inch

All have half covers for making large hole directly over fire. Water-backs furnished for each size, if ordered.

THE CORT CABOOSE.

A large, two-oven, caboose range, made to meet the requirements of very large vessels.

It is strongly and carefully made, and is warranted to work
perfectly.

The oven sides against the fire-box, if ever burned out, can be
easily taken out and replaced by loosening three bolts, without
disturbing any other plate.

The hinge-pins are all of Tobin Bronze, and cannot rust.

Half covers are furnished for making large boiler hole directly
over fire.

A warming closet is under each oven.

Top is surrounded by strong rail, removable at will.

Ends have rings for securing range to its place.

The range is fitted with flat grate ; or with triangular revolving
grate, as preferred.

Orders should distinctly say which style of grate is desired.

The range is constructed to use a water-back when desired,
which will be furnished if ordered.

THE CORT CABOOSE.

Size of Ovens	Size of Top	Number of Cover Holes	Size of Cover Holes	Hight of Range	Hight to top of Rail
13 x 19 x 11	3 ft. 11 in. x 2 ft.	6	8-inch	2 ft. 4 in.	2 ft. 9 in.

THE SHIPMATE.

The Shipmate is made specially for vessels. It is, in every way, suited to the peculiar requirements of vessel work, and to the convenience of its users.

It is thoroughly well made and durable, and is warranted to work perfectly.

The parts exposed to the fire are made especially heavy, so that repairs are seldom needed. The oven is large, with drop door. The fire-box and ash pit are also of ample size.

The hinge-pins are all of Tobin Bronze, and cannot rust.

The flues are large and deep, insuring perfect operation without need of frequent cleaning.

The top is surrounded by a strong rail, either part of which can be lifted out of its socket at will. The ends have rings to secure range to its place.

Two styles of top are made for this range : where the room will admit it, a large top, affording unusually large surface, is furnished (see cut, opposite page) ; where the space is scant, the range is fitted with a smaller one, as shown on page 64.

Two styles of feet are also made, one being two and one-half inches higher than the other.

Either style of feet can be used on any Shipmate Range.

THE SHIPMATE

With Large Top and High Feet.

	Size of Oven	Size of Large Top	Hight of Range	Hight to top of Rail
No. 7 .	16¼ x 16¼ x 10	26 x 33	25¼	29¼
No. 8 .	20 x 18 x 11	28 x 36	26¼	30¼

Orders should distinctly say whether this style of top, or small top as shown on next page, is wanted; and whether low feet or high feet are wanted.

No. 8 will be fitted with water-front, if ordered so.

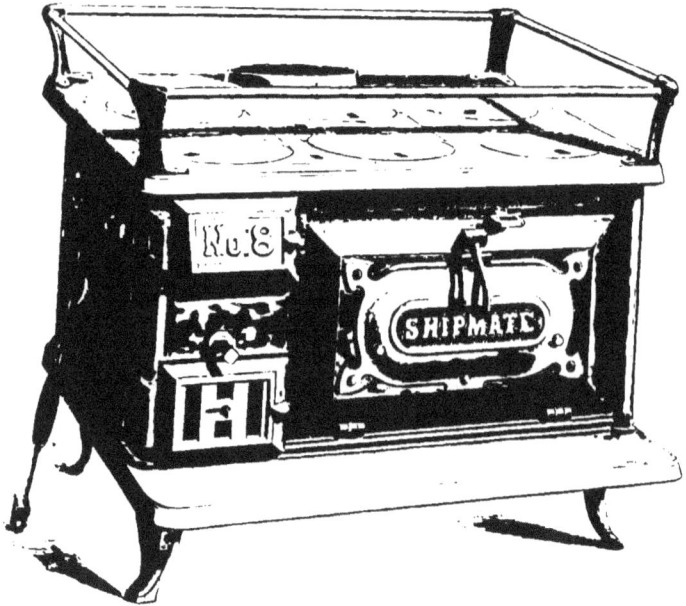

THE SHIPMATE

With Small Top and Short Feet.

	Size of Oven	Size of Small Top	Hight of Range	Hight to top of Rail
No. 7	16½ x 16½ x 10	24 x 30	23	27
No. 8	20 x 18 x 11	26 x 33	24	28

Orders should distinctly say whether this style of top, or large top as shown on page 63, is wanted; and whether low feet or high feet are wanted.

No. 8 will be fitted with water-front, if ordered so.

Memoranda.

THE YACHT.

What is said of the character of the larger cabooses is true of the Yacht, the difference being in the size.

The same care is used in its construction, and it is as well calculated to perform the work required of it.

It is especially suited to yachts, fishing smacks, lighters, small tugs, etc.

Has top rail, rings in end to secure it to its place ; oven door drops to front and forms shelf for basting meats and trying bread ; top is large ; grate and covers are heavy and durable.

Its operation is guaranteed.

For those who prefer to have the range higher than shown in cut, the range is fitted with high feet, raising it two and one-half inches higher than hight given in table of dimensions.

THE YACHT.

Four Holes.

	Size of Oven	Size of Top	Hight of Range	Hight to top of Rail
No. 7	14 x 11½ x 8½	23 x 24	20½	25
No. 8	16 x 11½ x 8½	24 x 25	20½	25

For No. 2, see next page.

THE YACHT.

Two Holes.

No. 2 has two 8-inch holes, and ring to make one 10-inch hole or, if desired, reducing rings for 7-inch covers.

Size of Oven	Size of Top	Hight of Range	Hight to Top of Rail
12 x 11½ x 8¼	24 x 16½	20¹	25

For larger sizes, see preceding page.

DECK IRONS

With Water Cups.

For pipe holes in vessels.   A protection against fire and water.

Plain or galvanized iron   4, 4½, 5, 5½, 6, and 6½ inches.

SQUARE BOILERS

With Round Sunk Pit.

See cut, page 63.

For 7-inch hole   .   .   .   .   9⅞ x 9⅞
For 8-inch hole   . 9¼ x 9¾, and 9¾ x 11¾

THE AMERICAN.

A Surface-Burning, Base-Heating Stove.

Of elegant proportions and design. It has deep oven, with kettle hole in bottom, and divided covers swinging in opposite directions, so that full depth and hight of oven can be utilized, and covers opened without interference with pipe or removal of urn; foot-rail firmly adjusted without bolts, and readily removed; a perfect anti-clinker grate, shaken or cleaned through poker-door in front, and tipped for removal of clinkers, from the outside, without opening ash-door; cast-iron fire-pot lined with brick; dust damper, use of which prevents any escape of ashes very useful also as a check-draft; deep bottom flues easily cleaned from the inside by removing plate full size of ash pit bottom.

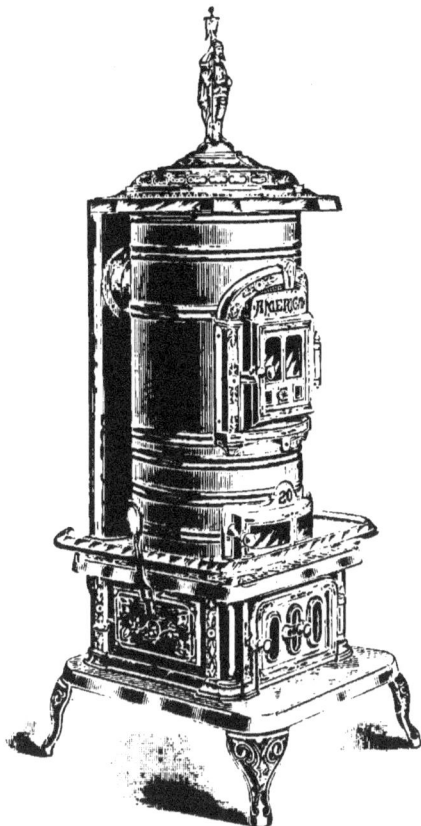

THE AMERICAN.

Round Stoves.

		Diameter of Fire-Pot
No. 10	.	. 10 inches
No. 15		11 inches
No. 20		12 inches
No. 30		15 inches
No. 40	. . . .	18 inches

Sets of castings with nickel foot-rail and top ring, or stoves mounted complete.

For oval stoves, see next page.

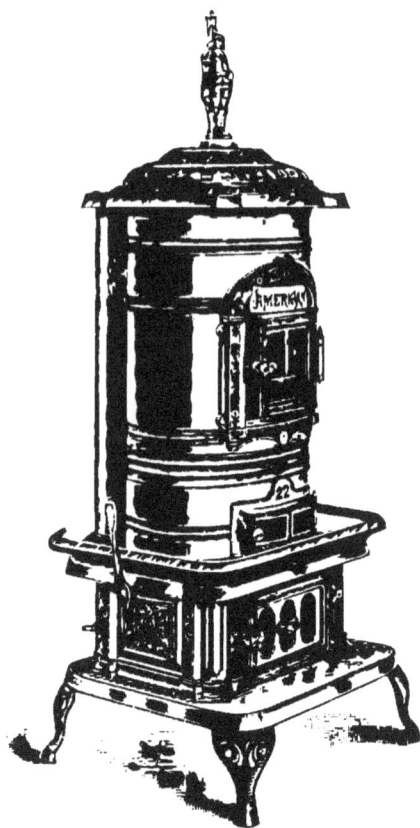

THE AMERICAN.

Oval Stoves.

			Diameter of Fire-Pot
No. 11	. .		10 inches
No. 22 .	. .	.	12 inches

Sets of castings with nickel foot-rail and top ring, or stoves mounted complete.

For round stoves, see preceding page.

Memoranda.

THE AMERICAN PARLOR HEATER.

A handsome, surface-burning, double-heater. Durable, easily managed, economical. While, if desired, all heat can be confined to the room in which the heater stands, unusually large radiating surface in the upper section, similar to that in the celebrated American Heater, gives great heating power for the room above, without serious diminution in that below.

This heater has the same lower construction and style as the American Parlor Stove, illustrated on page 71.

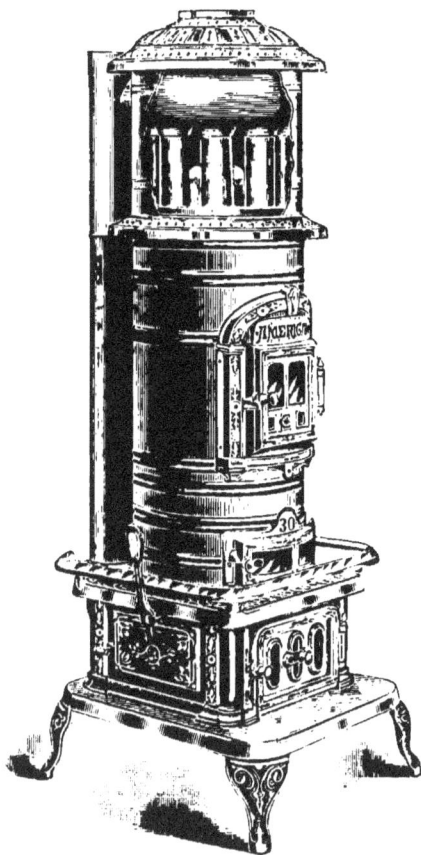

THE AMERICAN PARLOR HEATER.

	Diameter of Fire-Pot
No. 30	15 inches
No. 40	18 inches

Sets of castings, etc., or mounted complete.

THE LEDGER B CYLINDER.

The largest line of cylinder stoves in the market.

These cylinder stoves belong to the very highest class of their kind.

All sizes are provided with lifting cover beneath the swing cover; with mica in feed door; check-draft in feed door; toothed ring surrounding anti-clinker grate, which shakes and dumps from the side; a small drop door for the convenient removal of clinker without opening the swing hearth; and swing hearth fitted with a draft-slide, and so shaped that ashes cannot bank against it and scatter when it is opened.

Ash pit is deep and capacious.

Decorated porcelain bowl with receiver, as shown in first cut, furnished with the small sizes, or, for a small additional cost, a nickel receiver may be substituted for the usual one.

The larger sizes have a top ornament, as shown in second cut. Body rings for making higher stoves can be furnished, when ordered, for the four large sizes.

THE LEDGER B CYLINDER.

No. 8 No. 9 No. 10 No. 11 No. 12 No. 14 No. 16 No. 18 No. 20

Numbers indicate diameter in inches.
Sets of castings, or mounted complete.
For stoves with extension tops, see next page.

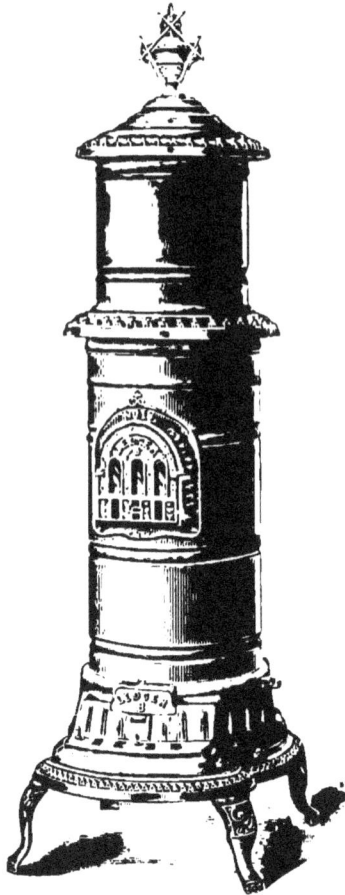

THE LEDGER B CYLINDER

With Extension Tops.

No. 14 No. 16 No. 18 No. 20

This cut shows form used for large sizes, when more radiating surface is required, with a better appearance than would be presented by carrying up the full size of body.

For the largest sizes, still another story is often added.

For stoves without extension tops, see preceding page.

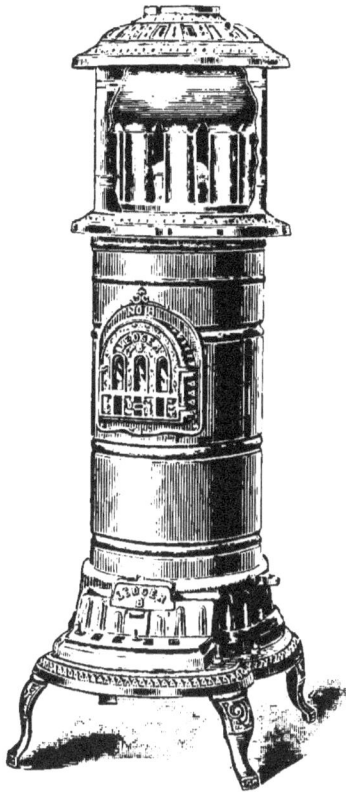

THE LEDGER B PARLOR HEATER.

No. 18. Diameter, 18 inches.

A highly efficient, low priced, double-heater, constructed from the No. 18 Ledger B Cylinder Stove. Very desirable where apparatus for heating two stories is wanted for a small expenditure of money.

THE LEDGER C.

Self-Feeding.

A powerful, durable, and low priced magazine stove.

It has a capacious feeder, easy of access through front feed door, leaving the flat top with cover hole free for heating, etc. Large illuminating doors give a cheerful appearance, and with fine proportions and tasteful design make the stove suitable for any office or store, while its great heating power renders it fit for depots, factories, etc.

Of the lower parts it need only be said that they are the same as those which have done such excellent service in the Ledger B Globe and Cylinder Stoves.

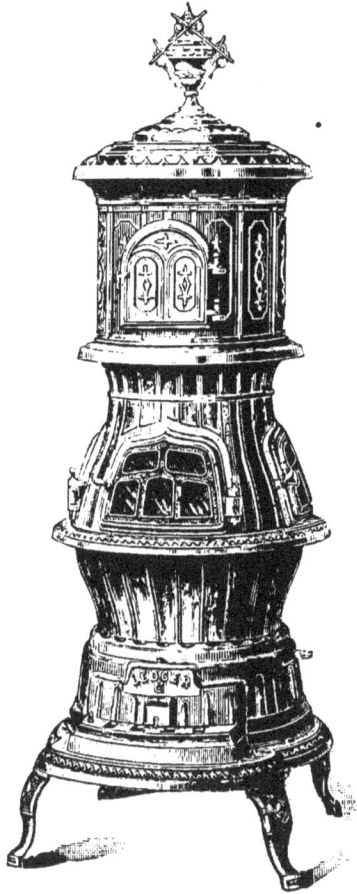

THE LEDGER C.

Self-Feeding.

No. 12 No. 14 No. 16 No. 18 No. 20

THE LEDGER B GLOBE STOVE.

Numbers indicate diameter in inches where fire-pot sits on the base.

	Diameter of Globe	Depth of Ash Pit Section	Hight on Feet
No. 7	9 inches	5 inches	26 inches
No. 8	10 "	5½ "	27 "
No. 9	11 "	6 "	31 "
No. 10	12 "	6½ "	32 "
No. 11	13 "	7 "	34 "
No. 12	14 "	7 "	36 "
No. 14	16	8 "	41 "
No. 16	18 "	9 "	44 "
No. 18	21 "	9½ "	48 "
No. 20	23 "	10 "	51 "

This remarkably large line of Globe Stoves, ranging, with but little difference between sizes, from seven inches to twenty inches, inclusive, covers every possible demand.

Made for situations where stoves are subjected to hard work and no favors are shown.

Specially suited to shops, stores, offices, schools, railway cars, or to any place where a quick and powerful heat is required.

Have convenient anti-clinker grate, as shown on page 84, large ash pit, swing hearth with draft-slide in it, small drop poker-door to allow cleaning grate without opening hearth. Register check in feed door.

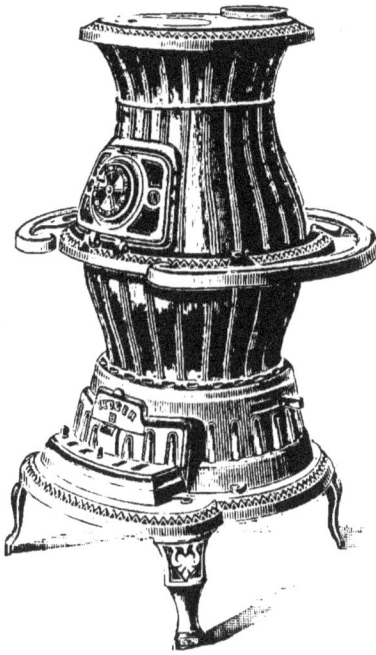

THE LEDGER B GLOBE STOVE.

For dimensions, etc., see opposite page.

No. 7 and No. 8 do not have rail.

No. 9, No. 10, and No. 11 can have top section adapted for sheet-iron drum, if so ordered.

All of the larger sizes have rings under the tops to support drum, as shown on next page, so that sheet-iron drum, taking top already on the stove, can be put on by simply loosening two bolts and raising the top.

THE LEDGER B GLOBE STOVE
With Sheet-Iron Drum.

For dimensions, etc., see page 82.

No. 12, No. 14, No. 16, No. 18, and No. 20, in this form, have the top and casting under the drum the same as those in the plain stove; so that the drum may be used or not, according to the situation.

No. 9, No. 10, and No. 11 have special top for drum.

The above cut shows the anti-clinker grate used in all of the Ledger B Stoves.

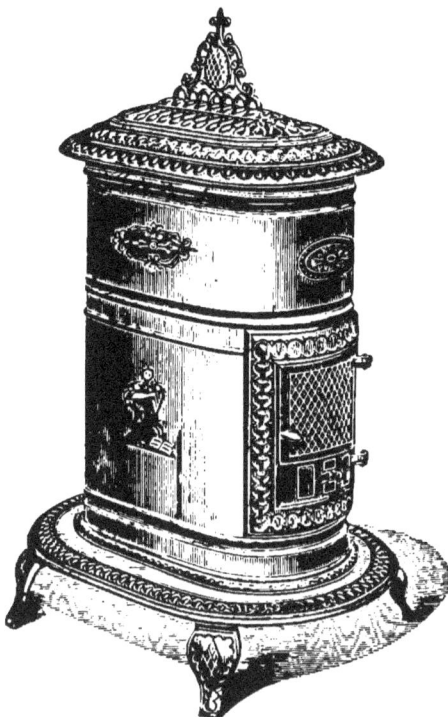

WOOD AIR-TIGHT.

Three Sizes.

With large oval opening under fret-work top, useful for boiling or heating.

Sets of Castings.

Numbers indicate the length of oval, in inches.

No. 18, No. 21, No. 24.

STAMFORD LAUNDRY STOVES.

Twenty Styles and Sizes.

For large laundries or small laundries, for hotels or private families, and for tailoring establishments.

Several of the cuts represent two or more sizes, a description of each size being given with cut representing its style.

The points of difference in the various styles give each a peculiar advantage for some particular use ; and while each is thoroughly well made and durable, the differences in size and form allow such an extensive range of prices that each buyer can readily select one suited to his means.

LAUNDRY D FURNACE.

For Hard Coal.

Brick Lined Fire Pot.

Cheap laundry and kitchen stove for all purposes excepting baking.

With patent rotary top, which forms at will either a direct connection with smoke-pipe, or a flue through which the flames must pass around the whole top before reaching its exit.

Size of cover hole, 9 inches.

Hight of stove, 13 inches.

OVAL TOP LAUNDRY FURNACE.

For Hard Coal.

Brick Lined Fire-Pot.

No. 7   . Two 7-inch holes.
No. 8   . Two 8-inch holes.

No. 7 has half covers to make 10-inch boiler hole.
No. 8 has half covers to make 11-inch boiler hole.
Hight of stoves, 14 inches.

No. 10 LAUNDRY STOVE.

This stove and No. 20, shown on next page, are also suitable for Confectioners' and Hatters' use.

Has 7-inch hole, and rings to make 8¼-inch and 13-inch holes.

Has a round sunk pan which fits the largest hole, and which has cover hole in centre and is useful for heating sad-irons, or for bringing boiling kettle close to fire. Has brick-lined fire-pot, deep ash pit, dumping and shaking grate, and swing hearth with draft-slide.

Hight of stove, 25 inches.

No. 20 LAUNDRY STOVE.

This stove and No. 10, shown on preceding page, are also suitable for Confectioners' and Hatters' use.

Has 9-inch hole, and rings to make 12½-inch and 17-inch holes.

Has a round sunk pan which fits the largest hole, and which has cover hole in centre and is useful for bringing boiling kettle close to fire. Has brick-lined fire-pot, deep ash pit, dumping and shaking grate, and swing hearth with draft-slide.

Hight of stove, 24 inches.

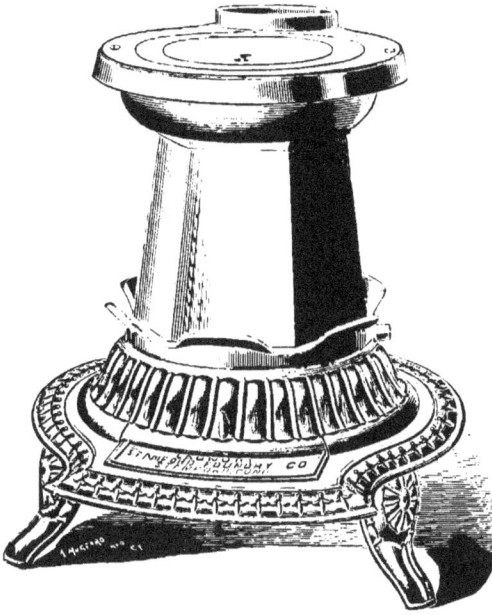

No. 61 LAUNDRY STOVE.

Takes six family irons on sides.   Has 7-inch hole, and ring to make 9-inch hole.

Large, deep ash pit.

Dumping grate easily removed and replaced, without disturbing any other part or removing any bolts or screws.

Diameter of grate	7½ inches
Hight of stove	19 inches

No. 62 LAUNDRY STOVE.

Takes six family irons on sides. Has two 8-inch holes, and centrepiece removable for using oval boiler.

Large, deep ash pit.

Dumping grate easily removed and replaced, without disturbing any other part or removing any bolts or screws.

Diameter of grate	7½ inches
Hight of stove	19 inches

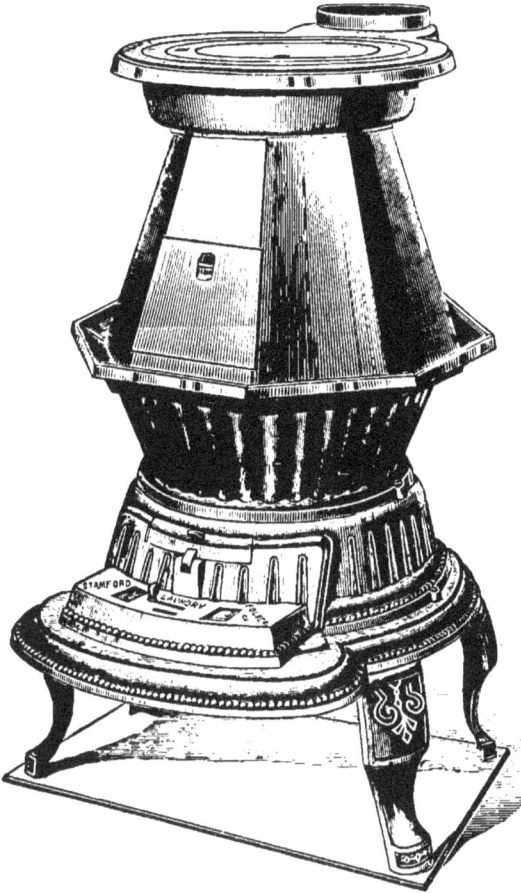

ROUND TOP LAUNDRY STOVES.

No. 6 takes six family, or six A, B, C, tailor irons on sides, and has ring in top to make 7-inch and 9-inch boiler holes.

No. 11 takes eight family, or eight A, B, C, tailor irons on sides, and has rings to make 8-inch, 10-inch, and 12-inch boiler holes.

No. 11 Goose Heater (no cut shown) has sides carried up to meet in pipe-collar, giving hight enough for longest tailors' geese, of which it takes eight.

These stoves have brick-lined fire-pot, deep ash pit, dumping and shaking grate, and swing hearth with draft-slide.

The drop feed door is hung without drilled hinges, so that there is no projection above surface to interfere with the irons.

Hight: No. 6, 29 inches; No. 11, 30½ inches; No. 11 G. H., 31 inches.

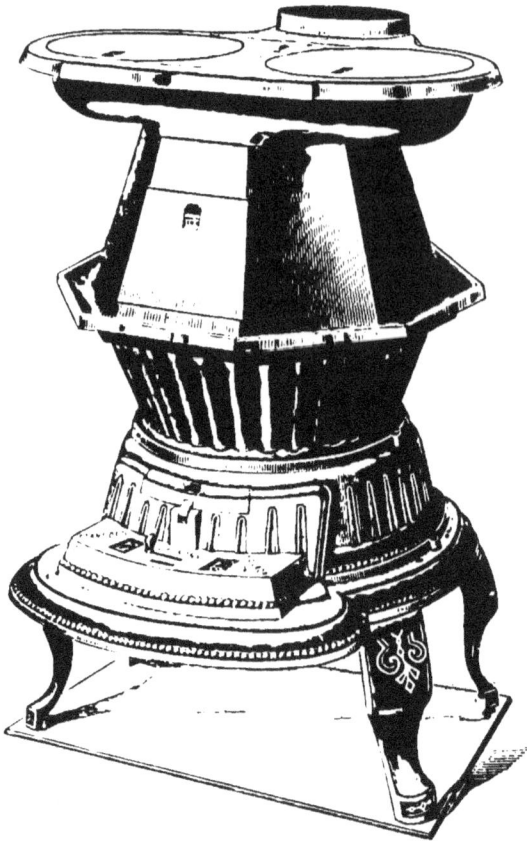

OVAL TOP LAUNDRY STOVES.

Two Sizes.

With two 8-inch holes, and centrepiece removable for using oval boiler.

No. 80 takes six family irons on sides.

No. 88 takes eight family irons on sides.

Both have brick-lined fire-pot, deep ash pit, dumping and shaking grate, and swing hearth with draft-slide.

The drop feed door is hung without drilled hinges, so that there is no projection above surface to interfere with the irons.

Hight of stoves: No. 80, 27 inches; No. 88, 29½ inches.

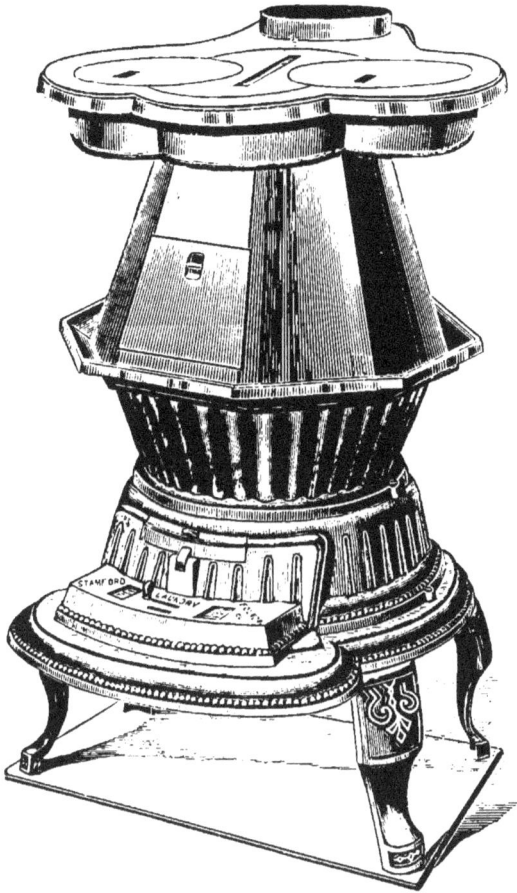

OVAL TOP LAUNDRY STOVES. Two Sizes.

For eight family, or eight A, B, C, tailor irons on sides.

No. 7 has two 7-inch holes, with half covers to make one 10-inch hole.

No. 8 has two 8-inch holes, with half covers to make one 11-inch hole.

Centres have deep flange on back end to compel heat to pass under cover holes.

Both stoves have brick-lined fire-pot, deep ash pit, dumping and shaking grate, and swing hearth with draft-slide.

The drop feed door is hung without drilled hinges, so that there is no projection above surface to interfere with the irons.

Hight of stoves, 31 inches.

LAUNDRY A.

Round Top, with Front Top Feed.

For eight family, or eight A, B, C, tailor irons on sides. Has rings to make 8-inch and 10-inch boiler holes.

Has brick-lined fire-pot, deep ash pit, dumping and shaking grate, and swing hearth with draft-slide.

Hight of stove, 31½ inches.

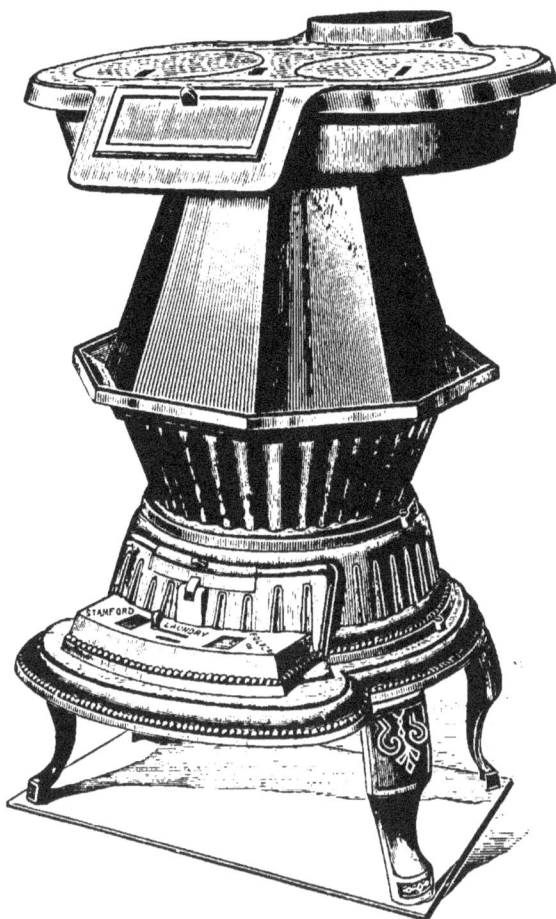

OVAL TOP, WITH FRONT TOP FEED.

Two Sizes.

For eight family, or eight A, B, C, tailor irons on sides.

B has two 7-inch holes, and centrepiece removable for using oval boiler.

C has two 8-inch holes, and centrepiece removable for using oval boiler.

Both sizes have brick-lined fire-pot, deep ash pit, dumping and shaking grate, and swing hearth with draft-slide.

Hight of stoves, 31½ inches.

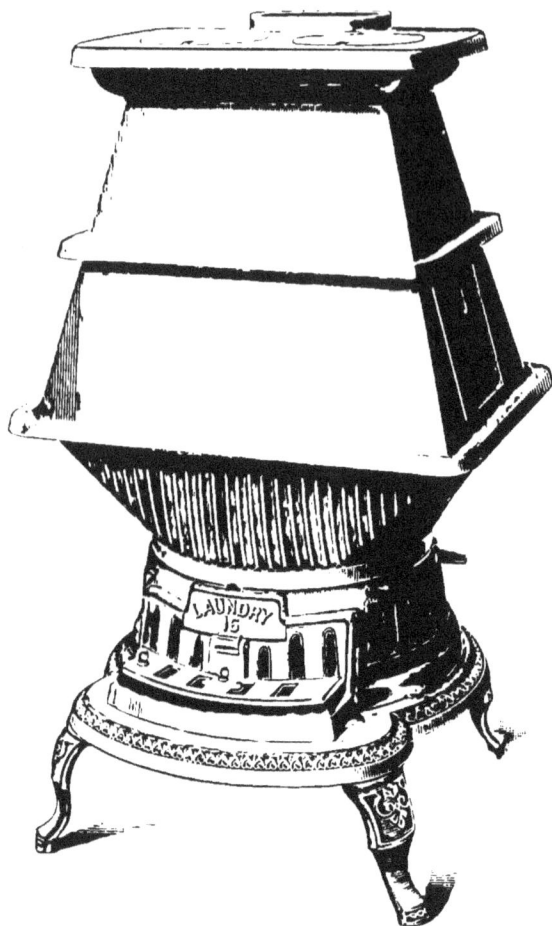

HOTEL LAUNDRY STOVE.
Two Sizes.

See opposite page.

For any situation where it is desired to heat a large number of irons.

No. 16. Two stories, for thirty A, B, C, tailor irons. Has two 7-inch boiler holes in top.

This stove is of heavy construction, has brick-lined fire-pot, and is fitted with shaking and dumping grate, and swing hearth with draft-slide and small poker-door.

Hight of stove, 44 inches.

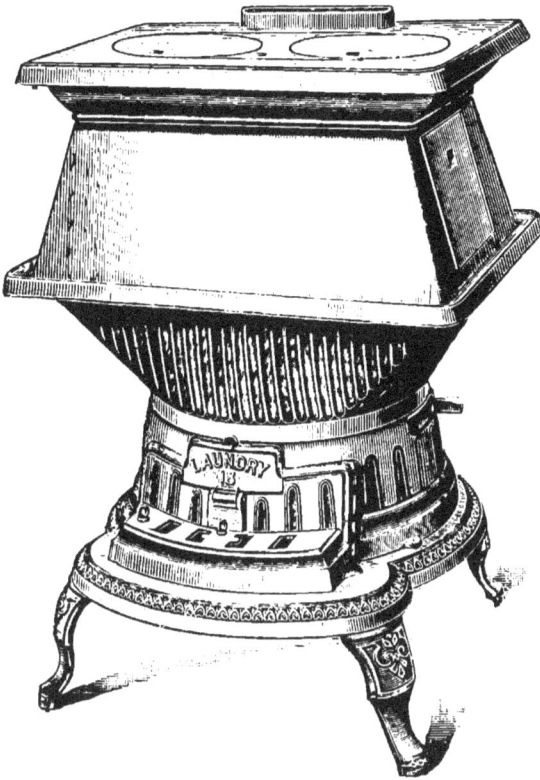

HOTEL LAUNDRY STOVE.
Two Sizes.
See opposite page.

For any situation where it is desired to heat a large number of irons.

No. 18. One story, for twenty A, B, C, tailor irons.

Has large top with two 9-inch boiler holes, and loose centre-piece, directly over and near to fire, giving great power for heating irons, or for boiling.

This stove is of heavy construction, has brick-lined fire-pot, and is fitted with shaking and dumping grate, and swing hearth with draft-slide and small poker-door.

Hight of stove, 36 inches.

TAILORS' GOOSE HEATER.

No. 17. For twenty-five geese, from 18 to 30 pounds. Has two 8-inch boiler holes, with loose centre, in top.

Is of heavy construction, has brick-lined fire-pot, and is fitted with shaking and dumping grate, and swing hearth with draft-slide and small poker-door.

Hight of stove, 39 inches.

Memoranda.

THE STAMFORD FURNACE.

No. 137   No. 142   No. 147   No. 157

This furnace is made specially low, to adapt it to use in very low cellars.

For dimensions, see page 105.

THE STAMFORD FURNACE.

FIRST SERIES.

Four sizes portable; three sizes for brick,

A good furnace must be durable, simple, economical. Some are economical and nothing else; others are simple and economical, but will not last.

The Stamford Furnace is one of the few that have all these essentials.

The BASE is strong and well braced.

The FIRE-POT is made very thick to stand the severest heat.

The DOME and RADIATOR are cast each in one solid piece, without joints; a feature important as regards both strength and safety. While these parts are thick enough to be free from any danger of injury from fire or rust, all unnecessary thickness is avoided, that the heat may the more readily pass through the shell.

The SMOKE PIPE is cast in one solid piece, and, extending through the casing, completes the working part of the furnace without the use of sheet-iron, which is so readily damaged by rust. This smoke-pipe can be taken out in any direction.

The Triangular Revolving Grate is recognized as the best furnace grate in use.

In the Stamford Furnace each pair of grate bars is worked independently of the other, so that a very small fire can be kept, when desired, by turning one pair of bars without disturbing the other pair.

The DOME affords space for burning the gases, and adds to the radiating surface.

The RADIATOR has a very large heating surface.

Its many upright pipes divide up the flames so that the heat is drawn out, and the flames are distributed instead of all going up one or two pipes.

The furnace is usually mounted with two casings, but is so made that three casings can be put on when desired.

It can easily be seen that the furnace is very simple; yet is perfectly complete.

Following are other important features:

IT CANNOT LEAK GAS. The few joints necessary have double cup-flanges, to be packed with sand or cement.

The FEED DOOR FRAME is attached to the cast-iron front with a slide joint, allowing for the expansion of the interior castings without any tendency to strain the front out of line.

The WIDE FEED DOOR allows feeding with a large shovel.

The DRAFT DOOR is so arranged that it may be worked by a chain from the rooms above.

The COLD-AIR CHECK-DRAFT on the smoke-pipe may also be operated from the floor above. The position of the opening is such as to prevent gas from escaping; yet the air is freely admitted.

The DUST FLUE creates a powerful suction from directly over the shaker opening, and thus prevents any escape of ashes. It may also be used as a check-draft.

The WATER PAN, independently supported by its own door-frame, can be placed where most convenient.

THE STAMFORD FURNACE.

Sets of castings are made in the following sizes :

FIRST SERIES : (see cut, page 102)

	Inside Diameter of Fire-Pot	Diameter of Casings	Extreme Hight of Castings
No. 137	20 in.	37 in.	4 ft. 5 in.
No. 142	22 in.	42 in.	4 ft. 7 in.
No. 147	23½ in.	47 in.	4 ft. 9 in.
No. 157	28 in.	57 in.	4 ft. 11 in.

SECOND SERIES : (See cut, page 106)

	Inside Diameter of Fire-Pot	Diameter of Casings	Extreme Hight of Castings
No. 237	20 in.	37 in.	5 ft.
No. 242	22 in.	42 in.	5 ft. 2 in.
No. 247	23½ in.	47 in.	5 ft. 4 in.
No. 257	28 in.	57 in.	5 ft. 6 in.

Casings can be furnished for all sizes.

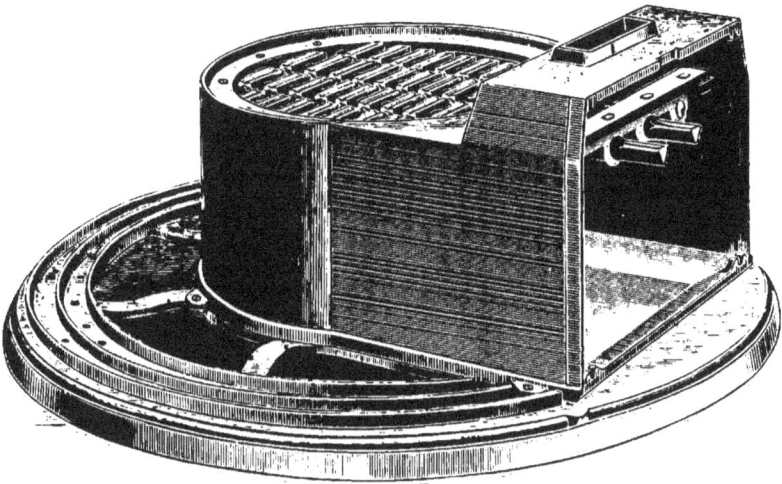

The above cut illustrates the triangular revolving grates used in all the STAMFORD FURNACES. They are strong and well fitted, and are so constructed that, should necessity for repairing ever arise, they can be easily removed by simply loosening two bolts immediately in front, without disturbing any other part of the furnace.

The two halves are worked independently, by separate shanks, so that when a small fire is desired, only one half of the grate surface need be cleaned.

THE STAMFORD FURNACE.

No. 237   No. 242   No. 247   No. 257

For dimensions, see preceding page.

THE STAMFORD FURNACE.

(See cut on opposite page.)

Besides the features enumerated in the general description of the Stamford Furnace, on preceding pages, this form has these peculiar ones :

In the unusually deep ash pit is a large SIFTING GRATE, which will be found very convenient for sifting the ashes before removing them from the furnace.

A hollow ring at the top of the fire-pot provides an ample supply of fresh, heated air at the very surface of the fire, to mingle with the gases and assist in completing their combustion.

If specially ordered so, this furnace can be fitted with flat, anti-clinker grate, such as is used in the Great American, instead of the triangular revolving grate ordinarily furnished.

THE GREAT AMERICAN HEATER.

For cut of this heater in brick setting, see page 111.

THE GREAT AMERICAN HEATER.

The Great American Heater is one of the most durable, practical, and successful heaters ever made.

It is of the highest grade.

It is self-cleaning, free from gas and dust ; not at all complicated in construction : any mechanic can set it.

It is popular because of economy, thorough efficiency, durability, and simplicity of management.

SMOKE PIPE can be taken out in any direction. It is cast in one solid piece, and extends out through the casing.

RADIATOR is cast in one section, without joints.

DOME is cast whole. It perfects combustion and increases radiation.

WIDE FEED DOOR allows feeding with a large shovel.

The durability of the FIRE-POT is attested by the fact that although large numbers of heaters, sold throughout many years, are still in constant use, the demand for pots for repairs is small.

ANTI-CLINKER GRATE is easily cleaned from outside without opening the door.

ASH PIT is deep.

The DUST FLUE creates a powerful suction from directly over the shaker opening, and thus prevents any escape of ashes. It may also be used as a check-draft.

WATER DOOR, with frame supporting the pan, can be placed at any hight on the furnace.

DRAFT-REGULATING DOOR may be operated from above.

COLD-AIR CHECK-DRAFT, attached to the smoke pipe, may be regulated from above.

This heater is SELF-CLEANING. Ashes cannot lodge in radiator section ; MUST FALL BACK into fire-pot.

IT CANNOT LEAK GAS. The few joints necessary have double cup-flanges, to be packed with sand or cement. This, together with the fact that the products of combustion have a direct ascent, and the furnace is self-cleaning, renders escape of gas impossible.

Unusually large surfaces for the distribution and radiation of heat insure great heating capacity and economy of fuel.

Sets of castings are made in the following sizes, the numbers indicating diameter of casings in inches :

Portable	For Brick Work	Inside Diameter of Fire-Pot	Extreme Hight of Castings
No. 23 .	——	. 15 in. .	4 ft. 5 in.
No. 28	——	. 17 in. . .	4 ft. 8 in.
No. 33	No. 33	19½ in.	. 4 ft. 10 in.
No. 38 .	No. 38	21 in. . .	4 ft. 10½ in.
No. 43 .	No. 43	23 in. .	5 ft. 1 in,
No. 53 .	No. 53	28 in. . .	5 ft. 3 in.

Casings can be furnished for all sizes.

Covering bars, and man-hole door and frame, can be furnished for all sizes for brick-work.

THE GREAT AMERICAN HEATER

In Brick Setting.

For description, see pages 109-110.

THE COMET FURNACE
With Steel Radiator.

THE COMET FURNACE

With Steel Radiator.

This furnace is as well made as anything can be.

It is not a cheap, slightly-built affair, made only to sell; it is made to do a great amount of heating, and to last as long as any other furnace under similar circumstances.

The radiator is made of No. 16 cold-rolled steel.

The cast-iron parts are in proportion.

The current of flame and smoke passes entirely around the radiator; cannot divide or go all on one side.

The dome rises nearly to top of radiator, and increases the heating surface.

The fire-pot is heavy and durable.

The Comet Furnace has the same perfect-working triangular grates as those used in the Stamford Furnaces. (See cut, page 105.)

The smoke-pipe can be taken out in any direction.

The drop draft-door and cold-air check-damper are made for regulating the fire from the rooms above.

The feed door frame is attached to the cast-iron front with a slide joint, allowing for the expansion of the interior castings without any tendency to strain the front out of line.

The wide feed door allows feeding with a large shovel.

The dust flue creates a powerful suction from directly over the shaker opening, and thus prevents any escape of ashes. It may also be used as a check-draft.

	Diameter of Fire-Pot	Diameter of Casings	Hight of Furnace less Casings
No. 132 . .	19½ in.	32 in.	4 ft. 7½ in.
No. 136 .	21 in.	36 in.	4 ft. 9 in.
No. 144 .	24 in.	44 in.	4 ft. 10¼ in.

Furnace can be furnished four inches lower, when necessary.

Casings can be furnished for all sizes.

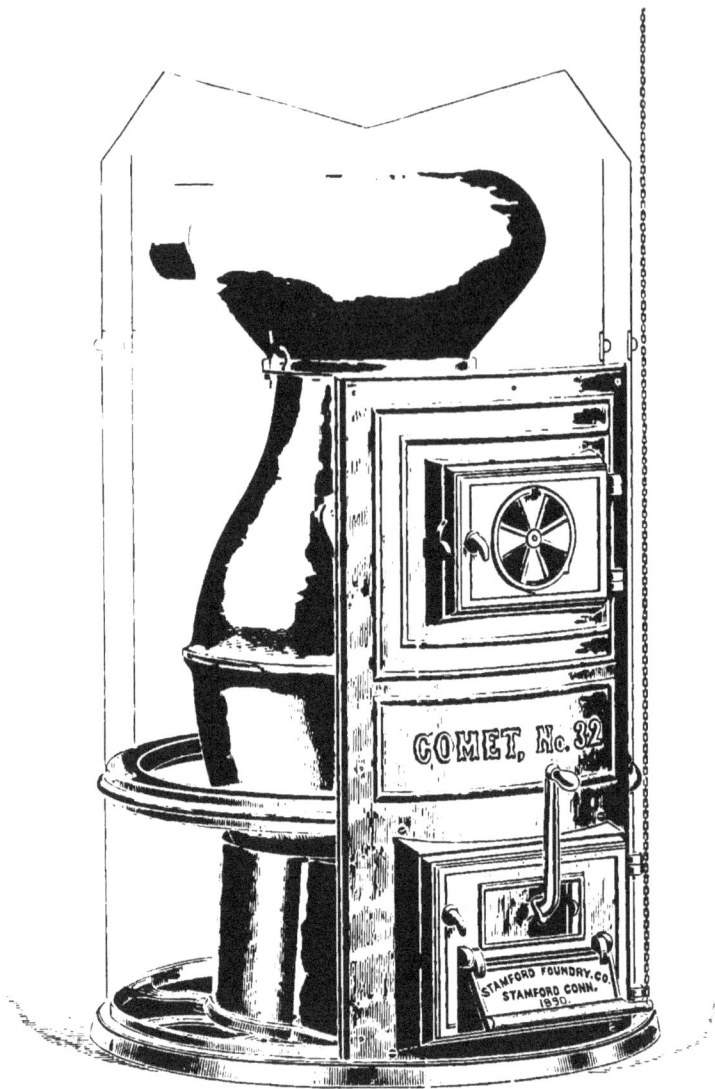

THE COMET FURNACE

With Cast Radiator.

No. 32   No. 36

THE COMET FURNACE

With Cast Radiator.

An important peculiarity of this furnace is its small hight : it sits so very low that it can be put into cellars where a person cannot even stand erect.

In its construction, durability has in no degree been sacrificed for the sake of cheapness. It is made heavy and strong to stand hard use.

There is nothing about it to wear out or fill up.

The shape of the radiator and body is such as to allow no chance for dirt to collect.

There is only one joint above the feed-mouth, and that is made with cup-ring for packing with sand or cement.

By an extension of smoke-pipe within the radiator, the heat is prevented from passing directly off, and is brought against all parts of the radiator.

The smoke-pipe can be taken out in any direction.

The drop draft-door and cold-air check-damper are made for regulating the fire from the rooms above.

This furnace has the same perfect-working triangular grate as that used in the Stamford Furnaces (see cut, page 105) ; or, if preferred and specially ordered so, it can have a plain anti-clinker grate, like that in the Great American.

	Diameter of Fire-Pot	Diameter of Casings	Total Hight of Castings
No. 32	19½ in.	32 in.	4 ft. 1 in.
No. 36	21 in.	36 in.	4 ft. 1½ in.

Casings can be furnished for both sizes.

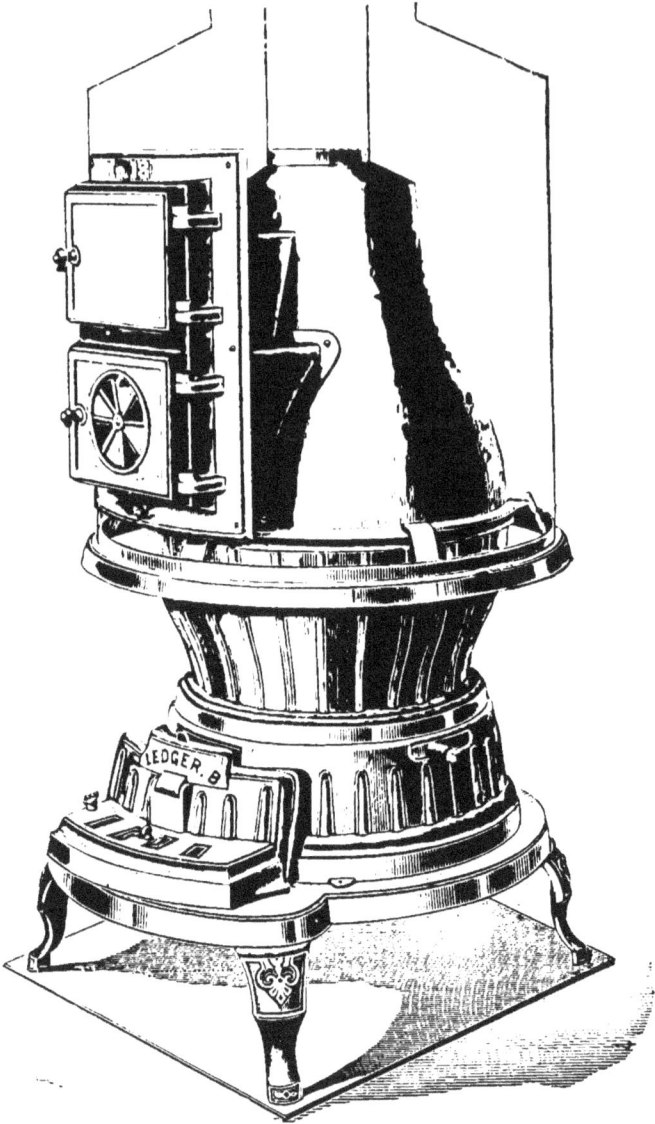

THE LEDGER B FURNACE.

THE LEDGER B FURNACE.

A Surface-Burning Double-Heater, for Heating Two or More Rooms on Different Stories.

This furnace is the well-known Ledger B Globe Stove, modified to support a casing for carrying a share of the heat to the room above.

Has convenient anti-clinker grate, large ash pit, draft-slide in ash-door, small drop poker door to allow cleaning the grate without opening the hearth ; register check in feed door, vapor pan with special door immediately over the feed door.

Body and fire-pot are heavy and strong, and entirely of cast-iron.

All parts are well made and carefully fitted.

When cased in Russia iron, the furnace is neat and attractive in appearance, and is just what is needed where a quick-acting, strong heat is desired for the room in which it stands, and for the floor above.

		Diameter of Casings	Inside Diameter of Fire-Pot
No. 16	.	24 in.	18 in.
No. 18	. .	26 in.	20 in.
No. 20	. .	28½ in.	22 in.

Casings can be furnished for all sizes.

Memoranda.

GENERAL